American Pearls

BY

HOWARD E. WASHBURN

British Library Cataloguing-in-Publication Data
A catalogue record for this book is available from
the British Library

AN INTRODUCTORY EXCERPT CHAPTER
BY
GEORGE FREDERICK KUNZ,

*"THE BOOK OF THE PEARL - THE HISTORY, ART, SCIENCE
AND INDUSTRY OF THE QUEEN OF THE GEMS" 1908*

PEARL-CULTURE AND PEARL-FARMING

Some asked how pearls did grow, and where.
Then spoke I to my girl,
To part her lips, and show them there
The quarelets of pearl.
HERRICK, *The Quarrie of Pearls.*

THE great profit that would accrue from an increased output of pearls has long directed attention to the problem of bringing this about by artificial means.

In his life of Apollonius of Tyana, Philostratus, a Greek writer of the third century, repeats a story afloat at the time, which credited the Arabs of the Red Sea with possessing some method of growing pearls artificially. The story as it reached Greece was that they first poured oil upon the sea for the purpose of calming the waves, and then dived down and caused the oysters to open their shells. Having effected this, they pricked the flesh with a sharp instrument and received the liquor which flowed from the wounds into suitable molds, and this liquor there hardened into the shape, color, and consistence of the natural gems.

While the description given by Philostratus is charged with many improbable details, and could scarcely develop belief, even in the most credulous, as to the exact method of procedure, it seems that the story may not have been wholly without foundation, and that attempts were made at that remote date to stimulate the growth of pearls.

In more modern times, the possibility of aiding or starting pearly formations in mollusks seems first to have been conceived by the Chinese about the fourteenth century. In 1736 there appeared in that storehouse of Oriental information, "Lettres édifiantes et curieuses écrites des missions étrangères,"[2] a communication from F. X. de Entrecolles, dated Pekin, 4th November, 1734, which set forth that there were people in China who busied themselves with growing pearls,

[1] Philostratus, "Vita Apollonii," *Lib.* III, c. 57, edit. Olearii, p. 139. Also see Konrad von Gessner, "Historiæ natura," *Lib.* IV, p.634.   [2] Vol. XXII, pp. 425-437.

and the product was not only vastly superior to the imitations manufactured in Europe, but were scarcely to be distinguished from the genuine. From Father Entrecolles's very detailed quotation of his unnamed Chinese authority, we condense this account. In a basin one half full of fresh water, place the largest mussels obtainable, set this basin in a secluded place where the dew may fall thereon, but where no female approaches, and neither the barking of dogs nor the crowing of chickens is to be heard. Pulverize some seed-pearls (*Yo tchu*), such as are commonly used in medicine, moisten this powder with juice expressed from leaves of a species of holly (*Che ta-kong lao*), and then roll the moistened powder into perfectly round pellets the size of a pea. These are permitted to dry under a moderate sunlight, and then are carefully inserted within the open shells of the mollusks. Each day for one hundred days the mussels are nourished with equal parts of powdered ginseng, china root, *peki*, which is a root more glutinous than isinglass, and of *pecho,* another medicinal root, all combined with honey and molded in the form of rice grains.

Although extremely detailed in some particulars, the Chinese account omits much to be desired as to the method in which the shells were opened to receive the pellets and the nourishment, and as to the importance of seclusion from females and loud noises. Admitting that it is "inaccurate and misleading," this letter seems to indicate very clearly that the Chinese had some method of assisting nature in growing pearls in river mussels.

The first person in Europe whose suggestion of the possibility of pearl-culture attracted general attention was Linnæus, the Swedish naturalist (1707–1778). In a letter to Von Haller, the Swiss anatomist, dated 13th September, 1748, he wrote: "At length I have ascertained the manner in which pearls originate and grow in shells; and in the course of five or six years I am able to produce, in any mother-of-pearl shell the size of one's hand, a pearl as large as the seed of the common vetch."[1] There was much secrecy about Linnæus's discovery, and even yet there is uncertainty as to the details of the method.

The Linnean Society of London apparently possesses some of the very pearls grown by Linnæus, as well as several manuscripts which throw much light on this subject. It appears from the latter that, under date of 6th February, 1761, Linnæus wrote that he "possessed the art" of impregnating mussels for pearl-production, and offered for a suitable reward from the state to publish the "secret" for the public use and benefit. A select committee of the state council of Sweden was appointed to confer with him, and on 27th July, 1761, the

[1] Pulteney, "General View of the Writings of Linnæus," London, 1805.

naturalist appeared and verbally explained his discovery. After various meetings, the select committee approved the "art" and recommended a compensation of 12,000 dalars (about $4800). It does not appear that the award was paid, and the following year the secret was purchased by Peter Bagge, a Gothenberg merchant, for the sum of 6000 dalars. On 7th September, 1762, King Adolph Frederick issued a grant to this merchant "to practice the art without interference or competition."

Peter Bagge was unable to exercise the rights which he had acquired, nor was he able to dispose of them to advantage. On his death the memorandum of the secret became lost, and it was not found until about 1821, when it was discovered by a grandson, J. P. Bagge. Under the date of 27th February, 1822, the King of Sweden confirmed to this grandson the privileges which his ancestor had purchased in 1762. Fruitless efforts were again made to dispose profitably of the rights either to individuals or to the Swedish government.

The details of Linnæus's "secret" have never been published authoritatively. In his "History of Inventions," Beckmann states that before the naturalist thought of the profits that might accrue from his discovery, he intimated the process in the sixth edition of his "Systema naturæ," wherein he states: "Margarita testæ excrescentia latere interiore, dum exterius latus perforatur."[2] "I once told him," says Beckmann, "that I had discovered his secret in his own writings; he seemed to be displeased, made no inquiry as to the passage, and changed the discourse."[3]

In the second volume of his edition of "Linnæus's Correspondence," Sir J. E. Smith remarks: "Specimens of pearls so produced by art in the *Mya margaritifera* are in the Linnean cabinet. The shell appears to have been pierced by flexible wires, the ends of which perhaps remain therein." Referring to this remark, J. P. Bagge comments: "This is the nearest I have seen any one come to truth, but still it will be remarked by reading the 'secret' that more information is required to enable persons to practice the art."

After a thorough examination of the manuscripts and other material, Professor Herdman concludes that the essential points of Linnæus's process are to make a very small hole in the shell and insert a round pellet of limestone fixed at the end of a fine silver wire, the hole being near the end of the shell so as to interfere only slightly with the mollusk, and the nucleus being kept free from the interior of the

[1] "Proceedings of the Linnean Society of London," October, 1905, p. 26.
[2] Pearl: an excrescence on the inside of a shell when the outside has been perforated.
[3] Beckmann, "History of Inventions," London, 1846, Vol. I, p. 263.
[4] London, 1821, p. 48.

shell so that the resulting pearl may not become adherent to it by a deposit of nacre.[1]

Shortly after Linnæus communicated with the Swedish government and before his death, it was learned in Europe that the art of producing "culture pearls" by a somewhat similar process had been practised by the Chinese for centuries.[2] They used several forms of matrices or nuclei, but principally spheres of nacre and bits of flat metal or molded lead, which were not infrequently in conventional outline of Buddha. In the spring or early summer, these were introduced under the mantle of the living mollusk after the shell had been carefully opened a fraction of an inch, and the animal was then returned to the pond or lake. The mollusk did its work in a leisurely way, like some people who have little to do, and many months elapsed before it was ready for opening and the removal of the pearly objects.

The most satisfactory description we have seen of this process appears to be that communicated nearly a century later to the London Society of Arts by Dr. D. T. Macgowan,[3] through H. B. M. plenipotentiary in China, from which this account is abridged and modified.

The industry is prosecuted in two villages near the city of Titsin, in the northern part of the province of Che-kiang, a silk-producing region. In May or June large specimens of the fresh-water mussels, *Dipsas plicatus*, are brought in baskets from Lake Tai-hu, about thirty miles distant. For recuperation from the journey, they are immersed in fresh water for a few days in bamboo cages, and are then ready to receive the matrices.

These nuclei are of various forms and materials, the most common being spherical beads of nacre, pellets of mud moistened with juice of camphor seeds, and especially thin leaden images, generally of Buddha in the usual sitting posture. In introducing these objects, the shell is gently opened with a spatula of bamboo or of pearl shell, and the mantle of the mollusk is carefully separated from one surface of the shell with a metal probe. The foreign bodies are then successively introduced at the point of a bifurcated bamboo stick, and placed, commonly in two parallel rows, upon the inner surface of the shell; a sufficient number having been placed on one valve, the operation is repeated on the other. As soon as released, the animal closes its shell, thus keeping the matrices in place. The mussels are then deposited one by one in canals or streams, or in ponds connected therewith, five or six inches apart, and where the depth is from two to five feet under water.

[1] "Proceedings of the Linnean Society of London," October, 1905, p. 29.
[2] See Grill, Abhandlungen der königlichen Schwedischen Akademie der Wissenschaften auf das Jahr 1772," Leipzig, Vol. XXXIV, pp. 88-90.
[3] "Journal of the Society of Arts," Vol. II, pp. 72-75.

If taken up within a few days and examined, the nuclei will be found attached to the shell by a membranous secretion; later this appears to be impregnated with calcareous matter, and finally layers of nacre are deposited around each nucleus, the process being analagous to the formation of calculary concretions in animals of higher development. A ridge generally extends from one pearly tumor to another, connecting them all together. Each month several tubs of night soil are thrown into the reservoir for the nourishment of the animals. Great care is taken to keep goat excretia from the water, as it is highly detrimental to the mussels, preventing the secretion of good nacre or even killing them if the quantity be sufficient. Persons inexperienced in the management lose ten or fifteen per cent. by deaths; others lose virtually none in a whole season.

In November, the mussels are removed from the water and opened, and the pearly masses are detached by means of a knife. If the matrix be of nacre, this is not removed; but the earthen and the metallic matrices are cut away, melted resin or white sealing-wax poured into the cavity, and the orifice covered with a piece of shell. These pearly formations have some of the luster and beauty of true pearls, and are furnished at a rate so cheap as to be procurable by almost any one. Most of them are purchased by jewelers, who set them in various personal ornaments, and especially in decorations for the hair. Those formed in the image of Buddha are used largely for amulets as well as for ornaments. They are about half an inch long, and while in the shell have a bluish tint, which disappears with removal of the matrix. Quantities of them are sold as talismans to pilgrims at the Buddhist shrines about Pooto and Hang-chau.

In some shells the culture pearls are permitted to remain by the Chinese growers, for sale as curios or souvenirs; specimens of these have found their way into many public and private collections of Europe and America. These shells are generally about seven inches long and four or five inches broad, and contain a double or triple row of pearls or images, as many as twenty-five of the former and sixteen of the latter to each valve. That the animal should survive the introduction of so many irritating bodies, and in such a brief period secrete a covering of nacre over them all, is certainly a striking physiological fact. Indeed, some naturalists have expressed strong doubts as to its possibility, supposing the forms were made to adhere to the shell by some composition; but the examination of living specimens in different stages of growth, with both valves studded with them, has fully demonstrated its truth.

It is represented that in the northern part of the Che-kiang province about five thousand families are employed in this work in connection
19

with rice-growing and silk-culture. To some of them it is the chief source of income, single families realizing as much as 300 silver dollars annually therefrom. In the village of Chung-kwan-o, the headquarters for culture pearls in China, a temple has been erected to the memory of the originator of this industry, Yu Shun Yang, who lived late in the thirteenth century, and was an ancestor of many persons now employed thereby.

The method in vogue in China for so many centuries has been the starting-point for similar attempts in various other countries. During the New Jersey pearling excitement in 1857, there were found several spherical pieces of nacre which had been introduced into Unios apparently for experimental pearl-culture; and in the collection of shells bequeathed to the United States National Museum by the late Dr. Isaac Lea, is a hemispherical piece of candle grease partly coated with pinkish nacre. Kelaart applied the Chinese method to the Ceylon pearl-oysters with much success in 1858. At the Berlin Fisheries Exhibition, in 1880, appeared the results of experiments in growing culture pearls in the river mussels in Saxony. Small foreign bodies had been introduced in the mantle, and others had been inserted between the mantle and the shell. These nuclei consisted of shell beads, unsightly pearls from other mussels, etc.; but unfortunately the shape of these was such that the mantle could not fit closely around them, consequently the result was so irregular as to be of no value except to show that German Unios as well as those of China could be made to cover foreign objects with pearly material.

Professor Herdman notes that, between 1751 and 1754, an inspector named Frederick Hedenberg received an annual salary "to inoculate the pearl mussels of Lulea (in the northern part of Sweden) with 'pearl-seeds' which he manufactured, and then to replant the mussels. Certain pearls were produced by the inspector, which it is recorded were sold for some 300 silver dollars."[1]

As noted by Broussonnet, in Finland artificial pearls were produced by inserting a round piece of nacre between the inner face of the shell and the mantle. The owner of the pearl fisheries at Vilshofen has succeeded in producing pearly figures by introducing into the mollusk flat figures of pewter, most of them representing fish in form.

In 1884, Bouchon-Brandely made experiments in pearl production at Tahiti. Gimlet holes about half an inch in diameter were drilled through different places in the shells of pearl-oysters, and through each of these holes a pellet of nacre or of glass was inserted and held by brass wire passing through a stopper of cork or burao wood, by means of which each opening was hermetically closed, so that the

[1] "Proceedings of the Linnean Society of London," October, 1905, p. 28.

pellet was the only foreign substance protruding on the inside of the shell.[1] The oysters were returned to the sea without further injury, and after the lapse of a month the pellets were found covered with thin layers of nacre.

Experiments in growing pearls in the abalone or Haliotis were made in 1897 by Louis Bouton, an account of which was given at the meeting of the Paris Académie des Sciences in 1898.[2] The tenacity of life in this mollusk makes it especially desirable for experiments of this nature. Through small holes bored into the shell, pellets of mother-of-pearl were inserted and placed within the mantle, the small holes being afterward closed up. Other nacreous pellets were introduced directly into the bronchial cavity. The objects were soon covered with thin, pearly layers, resulting in a few months in spheres of much beauty, resembling somewhat the pearls naturally produced by this mollusk. In six months, according to M. Bouton, the layers became of sufficient thickness to be attractive. Within limitations, the size of the pearl produced is in proportion to the length of time it is allowed to remain within the mollusk. The results of the experiments seem to encourage further efforts in this line, and possibly in course of time there may be a profitable business in growing pearls in abalones on the Pacific coast of the United States. Indeed, the experiments in transplanting and cultivating the pearl-oyster in Australia leads one to fancy that the culture of that species in the warm coastal waters of America is by no means an impossibility.

Many other experiments along similar lines have been made more recently. An interesting feature of attempts made by Mr. Vane Simmonds of Cedar Rapids, Iowa, in 1896–1898, is that in order to avoid straining the adductor muscles by forcibly opening the shell while the mollusk resisted the intrusion, each selected Unio was exposed in the open air and sunshine until the valves opened; then a wooden wedge was carefully inserted in the opening, and the mollusk immediately immersed in water to revive it or to sustain life. After a few moments of immersion, the operator carefully raised the mantle from the shell, inserted the pellet of wax or other small article to be covered with nacre, drew the mantle to its normal position, removed the wedge, and returned the mollusk to a selected place in the stream at sufficient depth to avoid danger of freezing in winter.

Probably it would be more satisfactory to stupefy the mollusks by means of some chemical in order to insert the pellets. Marine mollusks have been successfully stupefied by slowly adding magnesium sulphate crystals to the sea water until the animals no longer respond

[1] "La Pêche et la Culture des Huitres Per-lières à Tahiti," Paris, 1885.

[2] "Comptes Rendus de l'Académie des Sciences," Vol. CXXVII, pp. 828-830.

to contact. If treatment is not too prolonged, they may be returned to normal sea water with good prospects of recovery. To stupefy fresh-water mollusks, either chloral hydrate or chlorosone may be employed, although the latter is expensive to use in great quantity. Dr. Charles B. Davenport, of the Carnegie Institution, suggests that it might be well to experiment with pouring ether or chloroform over them.

In Japan the production of these pearly formations in *Margarit-ifera martensi,* which is closely related to the Ceylon oyster, has developed into some prominence since 1890, and the results have been well advertised. The industry is located in Ago Bay, near the celebrated temple of Ise in the province of Shima, and gives employment to about one hundred persons. It is stated that the proprietor, Kokichi Mikimoto, has leased about one thousand acres of sea bottom, on which are a million oysters of this species, which yield from 30,000 to 50,000 culture pearls annually.

As described by Dr. K. Mitsukuri, the shoal portions of this area are used for breeding the oysters and raising them to maturity, and in the deeper parts—covered by several fathoms of water—the oysters are specially treated for producing the culture pearls. In the former, the spat is collected on small stones, weighing six or eight pounds each, placed during May or June. The following November these stones, with the attached spat or young, are removed, for protection from cold, to depths greater than five or six feet, where they remain for about three years. At the end of that period, the growing oysters are taken from the water, the shells opened slightly, and rounded bits of pearl shell or nacre are introduced under the mantle without injury to the mollusks. About 300,000 are thus treated annually, and placed in the deeper water at the rate of about one to each square foot of bottom area. After the lapse of about four years more, the oysters are removed from the water and opened, when a large percentage of the pellets are found covered on the upper or exposed surface with nacre of good luster.

Most of these culture pearls are button-shaped and weigh two or three grains each. Although somewhat attractive and superior to the culture pearls of China and other fresh waters, they by no means compare favorably with choice pearls. They are rarely, if ever, spherical, and only the upper surface is lustrous; consequently they serve only the purpose of half-pearls. A cross section shows the nacreous growth in a thin concentric layer, forming a fragile hemispherical cap, the concave wall of which is covered with a brownish granular secretion which prevents perfect adhesion. Compared with choice pearls, they are not only deficient in luster, but are fragile, and are beautiful only on the upper surface, and not available for necklaces. Good specimens sell for several dollars each, and some individuals reach $50 or more. Specimens exhibited at the Paris Exposition in 1900 were awarded a silver medal; at the St. Petersburg Exhibition in 1902 they were awarded a gold medal; at the Tokio Exhibition a grand prize, and a medal at the St. Louis Exposition in 1904. The awards were given in the fisheries, and not the gem divisions.

The work of Mikimoto is not the only attempt now being made in Japan to produce pearls. A letter from Dr. T. Nishikawa, of the Tokio Imperial University, states: "It is a great pleasure for me to tell you that I am studying pearl formation and pearl-oyster culture in the university laboratory, and recently I have got my pearl laboratory at Fukura, on the Island of Awaji, where I began the pearl culture work this summer (1907). Fortunately, I found the cause of Japanese pearl formation, *i.e.*, the reason why and how the pearl is produced in the tissue of an oyster. I made practical application of this theory with great prospects for producing the natural and true pearls at will."

Among the most interesting of the pearl-culture enterprises are those of the Compañia Criadora de Concha y Perla, under the direction of Sr. Gaston J. Vives, in the Gulf of California. This company has an extensive station at San Gabriel, near La Paz, where breeding oysters are placed in prepared chests or cages for collecting the spat on trays. After remaining there for several weeks or months, the young mollusks are removed to prepared places (*viveros*) for further growth. Experiments are now made in depositing them between a series of parallel dams alternately touching each shore of a lagoon, thus developing a current of water over the oysters for conveying food to them, and thus hastening their growth.

In efforts to increase the output of pearls, attention has been given to the possibilities for extending the area and production of the reefs, and for stocking new areas and replenishing exhausted ones, thus bringing the pearl-bearing mollusks to maturity in greater abundance.

Although theoretically it does not seem a very difficult undertaking to cultivate the pearl-oysters by methods somewhat similar to the cultivation of edible oysters and clams, in no part of the world has this been successfully done on an extensive scale. While in certain minor cases, the areas of some species of pearl mollusks have been extended indirectly through man's agency—as the range of the Red Sea pearl-oyster into the Mediterranean since the Suez Canal was opened—there is no well-known instance in which new areas have been abundantly populated through direct efforts.

In the chapter on the pearl fisheries of Asia are noted the hitherto unsuccessful efforts made in Ceylon and India to preserve the young and immature oysters on the storm-swept reefs by removing them to less exposed areas. This has received close attention from the Ceylon authorities during the last two years. Other practical measures which are recommended for that region include "cultching," or the deposit of suitable solid material, such as shells or broken stone, to which the young oysters can attach themselves; thinning out over-crowded reefs, and cleaning the beds by means of a dredge, thereby removing starfish and other injurious animals. The attempts made by individuals and associations to extend the range of the reefs on the coast of Australia, among the Tuamotu Islands, in the Gulf of California, and some other localities, are noted in the appropriate chapters. But it may be stated that in most instances lack of adequate police protection has been not the least of the difficulties with which these experiments have had to contend.

Nor has much greater success followed upon efforts to prevent the exhaustion of the reefs and productive grounds through overfishing, except in those instances in which the government exercises a proprietory interest and determines the season, the area to be fished, and the quantity of mollusks to be removed. The most prominent instance of this is in Ceylon, where the fishery has been restricted to such seasons and periods as appeared to insure the maximum yield of pearls. Without restriction upon the fishery, the pearl-oyster in that populous region would doubtless become almost extinct in a few years. Another instance of proprietory interest on the part of the government is in some of the German States, where pearl fishing has been regulated and restricted for centuries. But there the sewage from cities and factories has accomplished almost as effectively, if less rapidly, what unrestricted fishing would have done.

Much attention has been given to the subject of pearl-culture in Bavaria, where the government has granted a small subsidy to encourage this industry, and a model pearl-mussel bank has been established in one of the brooks for the rational culture of the mussels.

On the Australian coast, the only theoretical protection of consequence is the restriction on taking small or immature oysters; but, owing to the great area over which the fisheries are prosecuted there, it has not been possible to enforce the regulations. At some of the Pacific islands and elsewhere, interdictions exist as to use of certain apparatus of capture, but this is intended for the purpose of reserving the industry to dependent natives rather than for protecting the reefs. Several efforts have been made to insure adequate protection for the Unios in our American rivers, but nothing in this direction has yet been accomplished by legislative enactment, except in Illinois.

Reference has already been made to the parasitic stage of Unios.[1] The attachment of the newly-hatched mollusks to the gills or fins of a fish is entirely a matter of chance, and unless this takes place they die within a few days. Under natural conditions the fish thus infected will rarely be found carrying as many of the parasitic Unios as they can without serious injury. If the fish are placed in a tank or a pond containing large numbers of newly-hatched Unios, it is possible to bring about the attachment of hundreds of them for every one that would be found there by chance of nature. A fish six inches in length may thus be made to carry several hundred parasitic Unios, and thus a thousand fish artificially infected may do the work of several hundred thousand in a state of nature. Experiments with small numbers of fish under observation in the laboratory indicate that their infection on a large scale is entirely possible, and the experiment by Messrs. Lefevre and Curtis now in progress at La Crosse, Wisconsin, in which over 25,000 young fish have been infected, gives every indication that such work may be begun even with the scanty knowledge now possessed.

Since it has already been shown that the production of pearls is an abnormal condition, it does not follow that an increase in the quantity of mollusks would necessarily result in a corresponding increase in the yield of pearls. Indeed, it might even be that the artificial conditions bringing about an enhanced prosperity and abundance of the mollusks would result in a corresponding decrease in the product of gems, the improved surroundings impairing if not destroying the conditions to which the pearls owe their origin. This has resulted in directing efforts toward abnormally increasing the abundance of pearls in a definite number of mollusks.

The development of the parasitic theory of pearl formation has naturally invited attention to the possibilities of increasing the yield of pearls by inoculating healthy mollusks with distomid parasites. It does not appear that this has yet advanced beyond the experimental stage, and virtually all that has been accomplished has been set forth in the chapter on the origin of pearls. It seems that there are great possibilities in the artificial production along these lines; and that under skilful management it could be made a profitable industry, especially if carried on concurrently with the systematic cultivation of mother-of-pearl shells.

Although there is scientific basis for the belief that it may be possible in time to bring about pearl growth in this manner, the public should not be too hasty in financing companies soliciting capital for establishing so-called "pearl farms." Every once in a while announcement is made in the public press of wonderful success which has been attained by some investigator, who surrounds his discovery with as much mystery as enveloped the Keeley motor, and who is as anxious to sell stock as was the owner of that mythical invention. A prospectus of one of these "pearl syndicates," which is now before us, claims to "increase and hasten pearl production by forcing the oyster, through doctoring the water in which it is immersed and also by irritating the mollusk itself." So far as the writers are aware, aside from the inexpensive but somewhat attractive culture pearls, no commercial success has yet followed the many attempts at artificial production.

This chapter should not close without reference to the so-called "breeding pearls," probably the most curious of all theories of pearl growth, regarding which many inquiries have been made. Throughout the Malay Archipelago there exists a generally accepted belief that if several selected pearls of good size are sealed in a box with a few grains of rice for nourishment they will increase in number as well as in size. If examined at the expiration of one year, small pearls may be found strewn about the bottom of the box, according to the theory; and in some instances the original pearls themselves will be found to have increased in size. If again inclosed for a further period of a year or more, the adherents of the theory say, the seed-pearls will further increase in size, and additional seed-pearls will form. Furthermore, the grains of rice will present the appearance of having been nibbled or as though a rodent had taken a bite in the end of each.

It is claimed that the breeding pearls are obtained from several species of mollusks, mostly from the Margaritifera, but also from the Tridacna (giant clam) and the Placuna (window shell). While cotton is the usual medium in which the pearls and rice are retained, some collectors substitute fresh water and yet others prefer salt water. It seems that rice is considered essential to success.

The earliest account we have seen of this extraordinary belief was given by Dr. Engelbert Kæmpfer,[1] who was connected with the Dutch embassy to Japan from 1690 to 1696, and since that time it has been referred to by many travelers in the Malay Archipelago.

A correspondent in the time-honored "Notes and Queries," 20th September, 1862, writes:

Nearly five years ago, while staying with friends in Pulo Penang (Straits of Malacca), I was shown by the wife of a prominent merchant five small pearls, which had increased and multiplied in her possession. She had set them aside for about 12 months in a small wooden box, packed in soft cotton and with half a dozen grains of common rice. On opening the box at the expiration of that time, she found four additional pearls, about the size of a

[1] Kæmpfer, "History of Japan," London, 1728, Vol. I, pp. 110–112.

small pinhead and of much beauty, which I saw and examined not long after the lady made the discovery. While my story may be received with laughter, I can most solemnly assure you of the truth of my having seen these pearls, and I have not the slightest doubt of the perfect truthfulness of the lady who possessed them. I questioned an eminent Malay merchant of Penang on this subject, and he assured me that one of his daughters had once possessed a similar growth of pearls. [1]

Notwithstanding the apparent absurdity of this pearl-breeding theory, belief in it appears to be not only sincere but wide-spread; as can be attested by any one familiar with affairs in the archipelago. A critical examination into the matter was made in 1877 by Dr. N. B. Dennys, curator of the Raffles Museum at Singapore, the result of which was communicated to the Straits branch of the Royal Asiatic Society, 28th February, 1878. [2] From his numerous quotations of persons who gave the results of their experiences we extract two instances. One gentleman had 120 small pearls in addition to the five breeding ones with which the experiment had started twenty years before, and during the entire period the box had not been molested except that it was opened occasionally for inspection by interested persons. Another experimentor inclosed three breeding pearls with a few grains of rice on 17th July, 1874; on opening the box on 14th July, 1875, nine additional pearls were discovered, and the three original ones appeared larger.

The belief has many curious variations. It is stated that in Borneo and the adjacent islands, many of the fishermen reserve every ninth pearl regardless of its size, and put the collection in a small bottle which is kept corked with a dead man's finger. According to Professor Kimmerly, nearly every burial-place along the Borneo coast has been desecrated in searching for "corks" for these bottles, and almost every hut has its dead-finger bottle, with from ten to fifty "breeding pearls" and twice that number of rice grains. [3] A correspondent at Sandakan, North Borneo, writes that at the time of his death at Hongkong in 1901, Dr. Dennys had in his possession a small box containing "breeding pearls"; but these disappeared after his death, and his brother, the crown solicitor, was unable to find them. This correspondent also states that the Ranee of Sarawak, a British protectorate in western Borneo, has a collection of "breeding pearls" numbering about two hundred, and that this is the only large collection known at present.

[1] "Notes and Queries," 3rd Series, Vol. II, p. 228.
[2] "Journal of the Straits Branch of the Royal Asiatic Society," Singapore, 1878, Vol. I, pp. 31-37.
[3] "Jewelers' Review," May 10, 1892.

As contrasted with abundant and unquestionably sincere testimony that pearls do "breed," it may be stated that absolutely no result has followed one or two native experiments made under supervision. While it must be admitted that negative evidence is always weaker than positive, and twenty failures would be outweighed by one successful experiment, yet the scientific objections to the possibility of pearls "breeding" cannot be overcome. The phenomenon is doubtless one of those curiosities of natural history in which some important factor has been overlooked.

Another curious theory is that peculiar pearls continue to grow after removal from the mollusk in which they originate. Quite recently it was reported from New Durham, North Carolina, that a pearl found there in 1896 had been growing continually since it was found and removed from the water. Unfortunately, it was weighed only when the last observation was made, and its increased size doubtless existed only in the imagination of its possessor.

CONTENTS

Plate   I.   Typical Pearl Shapes.

Plate II.   Specimens of Mussel Shells.

Plate III.   Mussel Shell and 68 grain Pearl.

Plate IV.   Necklace of Fine American Pearls.

Plate V.   Japanese Culture Pearls.

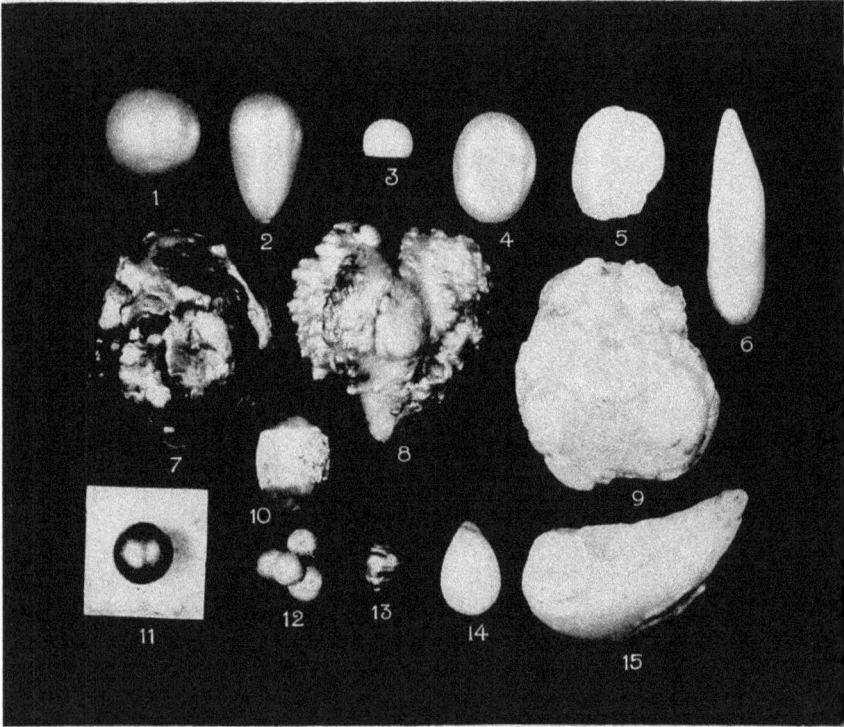

PLATE I. TYPICAL PEARL SHAPES, NATURAL SIZE.

Illustrated from fine American Pearls in the collection of Mr. Herma
Myer, 41 and 43 Maiden Lane, New York, N. Y.

PLATE I. DESCRIPTION AND WEIGHTS.

No. 1. Round pearl. 52 grains.

No. 2. Pear, (drop, oval or egg). 48 grains.

No. 3. Button pearl. 6 grains.

No. 4. Turtle-back. 32 grains.

No. 5. Flat Baroque pearl. 34 grains.

No. 6. Wing pearl. 18 grains.

No. 7. Dark green Baroque. 263 grains.

No. 8. Rare heart-shaped Baroque. Color—pink. 239 grains.

No. 9. Large Baroque. 308 grains.

No. 10. Banded pearl. Barrel shape. 31 grains.

No. 11. Round black pearl. 14 grains.

No. 12. Twinned pearls. Four have grown together. 11 grains.

No. 13. Pear shaped Baroque. Color—dark red. 9 grains.

No. 14. Flat pear shaped Baroque. 22 grains.

No. 15. Odd shaped Baroque. 108 grains.

AMERICAN PEARLS.

There are two classes of pearls: "oriental," and "fresh-water." In ancient times pearls were taken from oysters found in the seas along the coasts of Ceylon, Arabia and other Oriental countries. As the Orient was the source of pearl supply the pearls coming from there were called oriental pearls. The lustre of a pearl came to be called its orient, and the term is still used to describe the peculiar sheen of the gem. Although pearls are now found in other seas besides those of the Orient, the term "oriental pearls" is applied to all true pearls taken from oysters living in salt water.

At an early time, however, pearls taken from the fresh-water oysters were recognized as valuable gems. Pearls from the rivers of Britain were famous in the time of the Roman dominion. The streams of the British Isles and some upon the continent were the source of fresh-water pearls until the fisheries of America came to be worked. While these fisheries are still in their infancy they have given to the world a vast supply of pearls that rival the oriental pearls in beauty and value. These fresh-water gems taken from American streams and lakes are quite generally called American pearls.

When American pearls first came into the market there was a strong prejudice against them and many dealers thought they were of little value in comparison with oriental pearls. The following quotation from a jewelers' trade journal[1] is interesting chiefly as showing the attitude towards American pearls at that time.

"MILWAUKEE, WIS., Aug. 4.—Bunde & Upmeyer have on exhibition in their window a pearl taken from a Wisconsin clam which they claim is worth a large sum. They have made large purchases of Wisconsin pearls and no longer seek to keep secret the fact that they are dealing largely in these native gems.

"Mr. Upmeyer has been in Europe about a month, and has sold many Wisconsin pearls in that country. It is also stated that the firm has conducted a lucrative business with Tiffany & Co., New York."

"The other jewelers of Milwaukee, Wis., having made sport of Bunde & Upmeyer's statement, published above, and having declared that Wisconsin pearls have no value except as curiosities, a reporter for the New York *World* interviewed Charles F. Cook, of Tiffany & Co., in New York.

"Mr. Cook stated that his firm has several Wisconsin pearls valued at from $300 to $500 each, several Ohio pearls worth $900 each, and that he knows of single fresh water American pearls worth more than $2,000. He admitted that his house purchases fresh water pearls each year, and that the Wisconsin fisheries are the most prolific known."

It must be remembered that the above article was written at a time when the American fisheries were just beginning to be really worked. During the pre-

---

[1] The Jewelers' Weekly, August 7, 1890.

vious year (1889) large finds had been made along the Sugar River in Wisconsin and other states. Yet many jewelers thought that the gems were valuable chiefly as curiosities. Bunde & Upmeyer, however, recognized their value and bought a great many pearls, especially from the Wisconsin fisheries. At no time did this firm desire to keep secret the fact that they were "dealing largely in these native gems." They were among the first to buy American pearls and they still continue to deal in them extensively.

With the greater development of the American fisheries there has come a decided change in the regard for domestic pearls. While it is natural that wearers of pearls in America should have a preference for gems that are brought from afar, this preference is not so great but what the true beauty and worth of American pearls is generally recognized. In the words of a noted writer, "the fresh-water pearls may not surpass the oriental pearl but can without depreciation take their place beside it."

Perhaps one of the most distinctive characteristics of American pearls is their variety of colors. The brilliantly colored and richly lined mussels inhabiting American streams and lakes form pearls of almost every hue. They range in color from white, through pink, yellow, salmon, fawn, purple, red, green, brown, blue, black, passing through the several shades of these colors, often irridescent and of wonderful beauty. The irridescence adds to the lustre of the pearl, and some of the fresh-water gems are more lustrous than the best oriental pearls.

American pearls are also of many odd and curious shapes. Until quite recently these irregular shapes have not had a great deal of value. The fashion of eccentric designs in jewelry has created a demand for these baroque pearls because they can be used to excellent advantage in such work. Some of the finest of this baroque pearl jewelry is made by Crossman Co., of New York City. At one time only pearls of symmetrical shapes were sought, but baroques are now a prominent feature of the pearl industry and many persons deal in them exclusively.

The general price of pearls has been continually advancing and it is stated that this advance has been even more than that of diamonds in the last fifteen years. A Government Report[2] says: "The demand for American fresh-water pearls is strong, both in the domestic markets and abroad, especially in Paris, whither many pearls are taken directly from the pearl region." The readiness of foreign dealers to buy American pearls tends to raise prices here. The following extract is taken from a jeweler's trade journal:[3] "Next to the American buyers the French merchants have been most largely represented during the last season, buyers from houses in Paris having remained in the vicinity of the fisheries all the year. The activity of the Frenchmen in buying pearls of the coveted shapes, colors and quality, added much to the troubles of the New York bidders. It is esti-

---

[2] "The Production of Precious Stones in 1906, by Douglass B. Sterrett. Govt. Print, 1907.

[3] The Keystone, March, 1906.

mated that prices are now 25 to 35 per cent higher than they were a year ago, with the tendency continually upward." With the depletion of some of the larger American fisheries there has been reported a scarcity of good fresh-water gems which has tended to further increase their price.

It is well known that pearls have been prized as gems since the earliest records of mankind. From ancient times down to the present day the esteem for the pearl has not diminished. Rather it has increased. The subdued beauty and richness of a pearl appeals to all. Its costliness makes it a gem primarily for the rich. With the vast increase in the world's wealth during comparatively recent years, the use of pearls has been much extended. Demand for them is always great and people of wealth are willing to pay large sums for pearls that are the perfection of gems.

THE PEARL MUSSEL AND THE PEARL.

The Naiades, or pearl-bearing fresh-water mussels, commonly called clams, are found in many parts of the world. They are particularly abundant in the lakes and streams of the United States and Canada. It is estimated that there are over 600 distinct species living in North America, a large number of which are found in the Mississippi River and its tributaries.[4] The scientific names of these species are somewhat confused and are of little value except to the conchologist. Practically all of the varieties produce pearls.

The pearl is the one gem that is not a mineral. Since it is the product of animal life, our knowledge of it must come from a study of the animal that produces it. As yet complete knowledge regarding the pearl mussel or the pearl has not been obtained. Under the guidance of foreign governments elaborate investigations are being carried on at the marine fisheries to learn about the pearl oyster and its pearl. These investigations have proven valuable in enabling the fisheries to be regulated in a scientific manner, and also in giving much information regarding the formation of pearls. The importance of the pearl and pearl-button industries in the United States has made necessary a similar investigation here. It is

---

[4] Bulletin of the United States Fish Commission for 1898, p. 279.

said that the United States Fish Commission will undertake such an investigation in connection with the zoological department of the University of Missouri. The ultimate purpose will be to preserve the mussels as far as possible, and to devise a method to replenish the streams where the mussels have been recklessly exterminated.

The life of a mussel is one of the many marvelous chapters in the book of Nature. Many thousands of eggs are developed within the ovaries of a single female oyster. Each is a tiny bivalve when finally issuing from the parent mussel. As the little mussel lies upon the bottom of the stream or pond, a thread or filament floats up that will attach to any fish that comes in contact with it. When such an attachment is made the tiny clam draws itself up and fastens on to the fish, where it soon becomes covered over or encysted. In this abode the little mussel lives and develops for a period of about two months. It then falls to the bottom to meet with new vicissitudes and if fortune favors, to become a mature mussel. It is evident that the young of a mussel are subjected to a great many dangers. That one will be able to attach itself to a fish, is a remote probabiilty. There are other dangers, such as being destroyed by other animals. As a result only a small percentage of the young develop. Yet such myriads are hatched that the supply of mussels is usually numerous.

The mussel feeds by drawing a current of water through its body, gaining its sustenance from the ingredients of the water. In so doing it takes in many forms of animal life as well as inanimate particles.

Various kinds of mussels live under different conditions. Some are found in the quiet, almost stagnant water of ponds. Others live in the quiet, but fresher water of lakes. Some of the river mussels live in the deep and quietly flowing portions of the stream, while others are found in shallow, swiftly flowing water. There are also differences in the character of the river or lake bottoms in which the mussels live. Some varieties are found on mud, others on sand or gravel, and some exist among rocks and stones.

The mussel or clam moves about to a considerable extent. It does so by means of its foot, the coarse white part usually seen protruding from the shells when the mussel is undisturbed. During the cold seasons of the year, the mussels move to deep water. In the warm season they are inclined to find shallower places. This is particularly true in regard to lakes, and is of some consequence in streams, especially the larger rivers.

Most species of mussels are gregarious. Where there are any there are usually many. In the larger rivers these groups of clams constitute well defined beds. The size of some of these beds in the Mississippi River is very great. One was found near Muscatine, Iowa, about two miles long and a quarter of a mile wide. Before depleted by continued fishing, the mussels in these beds were densely crowded, not only covering the entire bottom but often several layers deep. Holes in the bottom were sometimes found to contain many bushels of mussels. While these are principally of two or three species, it is believed that there is a tendency for all mussels to be congregated. This does not mean, however, that mussels are not found scattered, because they are. But in most streams, certain portions prove to be more abundantly supplied with mussels than others.

The shells of a mussel consist of two valves which fit together and enclose the animal. In formation, each valve or shell consists of three distinct parts or strata. The epidermis is the rough exterior. It is called conchioline and is composed largely of animal matter. Next to this is a harder strata containing more carbonate of lime. On the inside and lining the shell is the pearly strata or nacre. This differs from the second in its construction. The second is formed of minute prisms arranged vertically to the surface, the third, or nacreus strata consists of minute layers parallel to the surface. These layers are finely folded or corrugated, and the pearly lustre is due to their action upon light.[5]

⌣ The mantle is the part of the mussel that builds the shell, creates the pearly lining and incidentally forms pearls. It is a thin, delicate membrane covering each valve and attached to it at the outer edge or pallial line. This organ by some mysterious process builds the shell in all its parts. It is supposed that different portions of the mantle deposit different shell materials. The extreme edge of the mantle is thought to excrete concholine, thereby continually building

[5] Encyclopedia Britanica, 9th Ed. 14:609.

out the edge of the shell. The surface next to the shell deposits the inner strata. If a shell is examined it will be noticed that certain parts of the pearly interior are more lustrous than others. The *posterior* end, or tip of a shell is always the brightest. (See Plates II and III). It is for this reason that pearls formed at this part of the mussel are the best, because given the finest lustre. The *anterior* part of the shell is a dull, opaque white, while the lower part usually has a good lustre, becoming brightest at the tip. It frequently happens that the interior of the same shell is of different colors, for instance, one end may be white and the other colored. It is not yet known what differences there are in the parts of a mantle to cause the various shell formations and colors, but that such differences exist is evident.

The mantle is the part of the mussel that creates the pearl. In secreting the material that builds the shell, it occasionally happens that concretions of these materials are formed apart from the shell. These are pearls. When these occur they may be composed entirely of the substance of any one layer of the shell, or of all the layers of which the shell is composed. Concretions of conchioline, the substance of the outer strata, occur, though rarely. They are usually of a dark brown color, often quite transparent, but they have little commercial value. Concretions of the substance of the second strata of the shell are most common. These are the chalky, lifeless pearls that are found so often but which have little value. Since a pearl is of no value as a gem unless it has a brilliant lustre, the gem pearl must be formed entirely, or at least externally of the lustrous inner portions of the shell. These concretions of the bright inner strata of the shell are the gem pearls, which are valued so highly.

The structure of a pearl consists of concentric layers similar to the structure of an onion. It frequently happens that the layers of a pearl are formed of different substances. The outer layer may be bright while the interior layers are chalky, or vice versa. When the outer layer is dull, it is sometimes peeled, in hope that the layer beneath may have a good lustre.

About 90% of the substance of a pearl consists of carbonate of lime, the remaining substance being organic matter and water. Since so great a portion of a pearl consists of carbonate of lime, it is thought that the presence of limestone rock along the course of the stream may have an influence upon the production of pearls in that more lime is carried by the water, thus supplying the mussel with more pearl producing material.

Pearls are formed in various shapes. The principal shapes are illustrated in Plate I, from specimens of fine American pearls.

Round pearls, and Pear Shapes (including Drop, Oval and Egg shapes), are formed within the soft portions of the mussel and are not attached to the shell.

Button pearls are those having a flat side, usually caused by attachment to the shell. Wing pearls are long, slender pieces, usually formed near the hing of the shell.

Baroques are irregular and odd-shaped pearls. Frank Koeckeritz, an experienced dealer in American pearls, estimates the proportionate occurrence of these shapes to be about as follows:

Drop, 5 per cent; Pear, 5 per cent; Oval, 10 per cent; Round, 15 per cent; Button, 25 per cent; Irregular, 40 per cent.

During the progress of pearl formation it frequently happens that two or more pearls become fastened together. These curious formations are called "twined" pearls. The surface of a pearl is often marred by points or knobs, as if sand or other foreign particles had been attached and covered with the pearly coating leaving the exterior of the pearl rough.

Pearls of good shape are frequently encircled by a narrow lustreless band, the other portions of the pearl being of good lustre. A band is a serious defect in a pearl and reduces its value very much. It is thought by some that "free" pearls (those unattached to the shell) are rolled or revolved while within the mussel. If this is true, it may account for the occurrence of banded pearls.

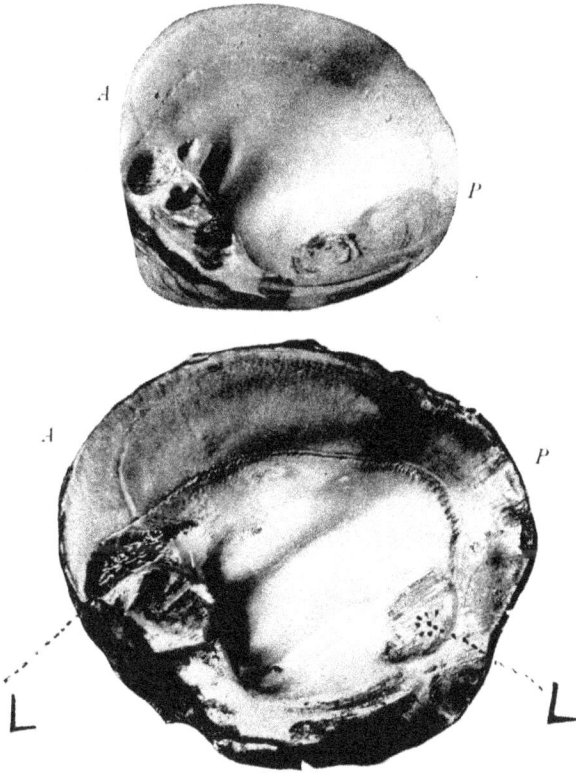

PLATE II. SPECIMENS OF MUSSEL SHELLS.
A—Anterior end.   P—Posterior end or tip.
L L—Ligaments.

AMERICAN PEARL FISHERIES.

The taking of pearls from American streams is not a new thing. The early explorers of the country found the native Indians in possession of a great number of pearls. In fact one of the incentives for many of the early conquests was the hope of obtaining pearls. As the Indians used the mussels for food it was natural that they should find many gems. The natives' attitude toward these, however, was quite different from that of the whites. While they no doubt appreciated in a way the beauty of the pearls, they ruined the gems by drilling them with heated instruments, thus rendering them valueless in the eyes of the new-comers.

After the Indians gave way to the white race the latter was so engrossed with the other natural resources of the new country that for a long time little attention was paid to the treasures of pearl contained in the numerous streams and lakes. It was not until the year 1857 that interest in American pearls was aroused. In that year the famous pearl known as the "Queen" was found at Notch Brook, New Jersey. This pearl weighed 93 grains, was perfect in every respect and was sold abroad for $2,500. It is now valued at several times that amount.

The discovery of the "Queen" pearl created a great deal of excitement. People abandoned their ordinary occupations and flocked to the streams to fish for pearls. While many pearls were found, the mussels were soon rendered so few in number that the work ceased to be profitable.

The history of pearl fishing in America, till quite recently, is but a repetition of pearl excitements in different parts of the country. The interest of a community is aroused by the finding of a valuable pearl and pearl fishing is then carried to an excess, with the result that the mussel beds are soon depleted. With the depletion of the mussel beds, the general interest in pearl fishing which has been aroused by the temporary excitement ceases, and the real work of pearl fishing is done by those who have the inclination and perseverance to carry it on.

The Little Miami River in Ohio is mentioned as one of the streams in which fine fresh-water pearls were found at an early date. Extended pearl fishing began in this river with the finding of several valuable pearls near Waynesville, Ohio, in 1876. This and other Ohio streams continue to yield many gem pearls.

In a few years the interest extended to other states. In 1889, the Wisconsin streams began to be fished for pearls. The excitement started when valuable finds were made in the Rock River and its tributaries. The pearls taken from the Sugar River are justly famous as ranking among the finest of American pearls. Although this stream has been overfished for many years, it still continues to yield valuable pearls. The activity in pearl hunting extended to many other Wisconsin streams as well as streams in Indiana, Illinois and other Northwestern States.

In the year 1897, there occurred a very great pearl excitement which centered in Arkansas but which influenced pearl fishing in many other states. It started by the finding of many pearls lying in the mud in the lakes and bayous of Western Arkansas. A Memphis Syndicate was formed and after some difficulties in the courts secured the exclusive right to dredge for pearls in two of the most favorable lakes. The newspapers gave much prominence to the pearl finds and these elaborate accounts aroused a general interest in pearl hunting. The streams of Arkansas as well as those of other states were for a while the scenes of immense activity. This excitement, however, quieted down in time, but pearl fishing continued to be carried on as an ordinary and profitable industry.

These are the principal steps in the development of American pearl fisheries. There are of course a great many instances of pearl finds having been made which attracted only the attention of the immediate locality. The pearl excitements mentioned directed the attention of nearly the entire people to the pearl resources of the country. As a result, a great number of fine pearls have been obtained which have established the importance of the American pearl fisheries as a source of pearl supply. Moreover, a permanent interest in the pearl fisheries has been established, and several thousand persons spend a part of each year in the work of gathering pearls.

The center of the pearl industry in the United States is undoubtedly along the Mississippi River. The pearl button factories furnish a ready market for the shells, which makes the taking of mussels a paying occupation even without the finding of pearls. As such enormous quantities of mussels are taken from this river each year, the output of pearls is very great. Some of the large rivers as well as smaller streams tributary to the Mississippi have the pearl-button industry as an additional stimulus to the pearl industry.

The Arkansas rivers, especially the White, Black, Cache, St. Francis, Red, Arkansas and their tributaries are the source of many of the finest American pearls. The Iowa, Cedar, Skunk, and Des Moines rivers in Iowa are fished to a considerable extent. The Wisconsin River is fished for pearls as well as the Rock River in Wisconsin and Illinois. The streams of Missouri, particularly the many tributaries of the Missouri River, are the source of many pearls. The rivers and streams of Ohio, Kentucky and Tennessee continue to produce some of the most valuable Amer-an pearls. The Ohio River has proved to be rich in s supply of pearls. Many valuable pearls have been und recently in the Kankakee River in Indiana and linois. Pearl fishing is an important industry along ..e Wabash River and its tributaries, the industry being centered about Vincennes, Indiana.

Pearl fishing is carried on in many streams besides those mentioned. The small rivers and creeks frequently contain the finest varieties of mussels and produce pearls of the best quality. These small streams are so numerous that the field for the pearl industry is very great. Many pearl hunters leave the grounds that are regularly fished and work in the smaller streams that have not yet been touched. A great deal of pearl fishing on these streams is done quietly, and the sources of pearl supply are often kept as secret as possible. A reecnt writer[6] states that the streams in the northwestern section of New York State are regularly fished, but without excitement. The Canadian streams are fished to quite an extent and it is believed by some that the northern streams contain the better varieties of pearl producing mussels.

It is practically impossible to determine with any degree of certainty the value of the output of pearls from the American fisheries. A Government Report[7] says: "The pearl industry is carried on in such a way that it is not possible to collect statistics showing the production. Buyers and dealers, not only from New York and other eastern cities, but even from Paris, visit the Mississippi region in the pearl-gathering season, travel from point to point, and at the end of the season return to their places of business. Many small dealers sell to larger ones on the spot; others send their product off to be marketed. In many cases parcels of pearls change hands two or three times before appearing in the gem markets. Pearls amounting to many thousands of dollars in value are exported annually, which apparently have not been reported to the Bureau of Statistics of the Department

---

[6] The Pearl (1907), by W. R. Cattelle, page 261.

[7] Production of Precious Stones in 1906, by Douglass B. Sterrett. Govt. Print.

of Commerce and Labor. Another part of the report reads, "the demand for domestic fresh-water pearls has been strong, and the production large, but it cannot be accurately given. They came chiefly from the Mississippi Valley region. A portion of the production has been exported, and in return there has been an importation of nearly two and a half million dollars worth."

The value of the American mussel fisheries has directed the attention of the government to the necessity of protecting them from exhaustion. In 1894 Mr. George F. Kunz, the noted gem expert, at the request of the United States Fish Commission, undertook an examination of the pearl industry in the United States. In his elaborate report on "The Fresh-Water Pearl Fisheries of the United States,"[8] Mr. Kunz carefully reviewed the conditions of the industry and recommended legislative protection to conserve the pearl resources. In 1898 Mr. Hugh M. Smith associated with the United States Fish Commission, examined the mussel fisheries of the Mississippi River and in his report[9] recommended their protection. One important suggestion made was to prohibit fishing during the portion of the year in which the mussels spawn. This is from January to May, though it is known that certain species spawn during the summer and fall. Other suggestions are to pre-

---

[8] United States Fish Commission Bulletin for 1897, pages 373 to 426. Plate I to XXII.

[9] The Mussel Fishery and Pearl-Button Industry of the Mississippi River. United States Fish Commission Bulletin for 1898. Pages 289-314. Plates 65 to 85.

vent the destruction of mussels from the pollution of streams by the refuse of factories and the sewage of cities, and to prohibit the taking of small mussels. In 1903 the state of Arkansas passed a law regulating the taking of mussels. The law prohibited mussel fishing from April 1st to June 1st, and also prohibited the use of the "crowfoot." When the law took effect in March, 1903, it created such an opposition from the mussel fishermen, the pearl buyers and button manufacturers that it was repealed on the 20th of May of the same year.

Mr. W. D. Burd, a pearl dealer who has had long experience with American pearls and fisheries, believes that the use of the crowfoot should be regulated. Many mussels are struck by the hooks and injured or killed but are not caught. Mr. Burd suggests that the hooks should be made of heavier wire and the prongs shorter and not so sharp.

At present there seems to be no regulation of the American fisheries and the fishermen are free to take the mussels at any time of the year, in any way and in any quantities even to the complete exhaustion of the beds. Recent agitations indicate that something may be done in the near future to regulate mussel fishing. Foreign governments controlling the marine fisheries secure careful investigation of the conditions of the fisheries in order to regulate them in a scientific manner. When more exact knowledge regarding the pearl mussel is obtained it is probable that much may be done to maintain and aid both the pearl and the pearl button industries.

PEARL FISHING.

In fishing for pearls, the first thing, of course, is to get the mussels. This is usually not a difficult matter. A few suggestions, however, and a brief description of some of the methods used in river fishing, may be of help to those trying the work for the first time, and along unfished streams where there is as yet no opportunity to see how the actual work is done.

We have already seen that mussels live in different portions of a stream and in water of various depths. It is necessary to try the stream or lake to find where the mussels are, and their situation will determine the best method for securing them. The implements principally used are the "water-telescope," the "hand-rake," and the "crow-foot."

The water-telescope is often used in shallow water, especially where the bottom is stony. The telescope is not difficult to make. It is merely a long narrow box made of thin boards. A convenient size is 6 inches wide, 4 inches deep, and 30 inches long. The bottom is fitted with a piece of glass and should be watertight. A handle may be put on for convenience. With this the fisherman wades in the water and submerging the glass end, is able to see plainly any mussels lying on the bottom. These can then be taken with a stick with a clasp on the end, or with a small rake. The water telescope can also be used from a boat in deep water when the water is not muddy. Another

PLATE III. MUSSEL SHELL AND 68 GRAIN PEARL.
A—Anterior end. P—Posterior end.

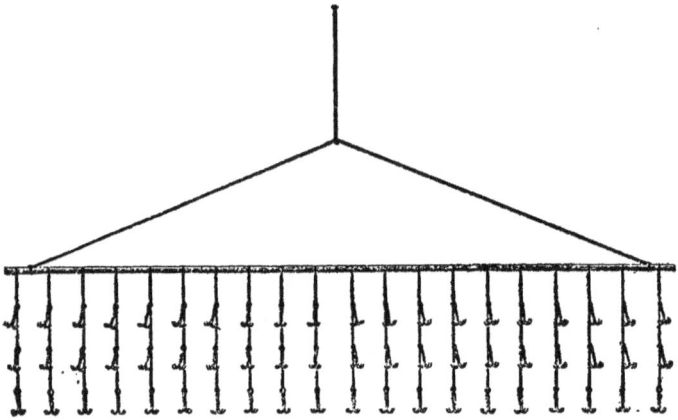

OUTLINE SHAPE OF ''CROW FOOT.''

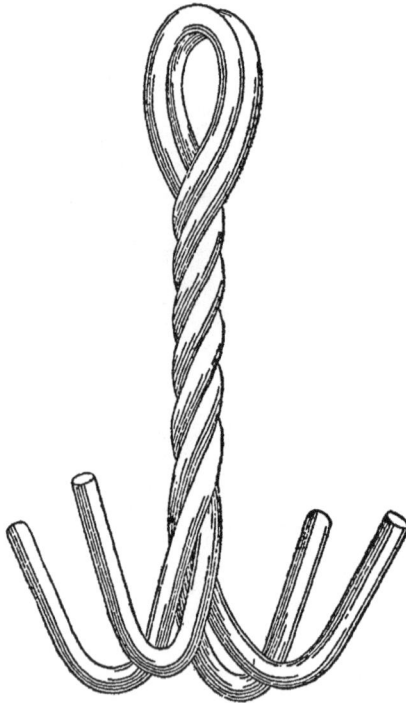

''CROW FOOT'' HOOK, ACTUAL SIZE.

form of the telescope is to have it made out of zinc with the glass end fitted in so as to be perfectly water tight.

The hand-rake is the usual implement used in taking mussels from the ordinary stream. This is a long handled rake with teeth 6 or 8 inches in length. The back of the rake is covered with wire netting forming a basket in which to receive the mussels. If the stream is wide and deep, the rake can be used from a boat; if the stream is narrow it can be used from the shore. The hand-rake is a common and effective means of getting mussels.

Another device that is used to a great extent along the larger rivers and especially the Mississippi, is the crow-foot. This consists of an iron pipe or bar about 6 feet in length, to which are attached, at intervals of about six inches, series of four-pronged hooks. The hooks are made of stout wire (about size No. 9), and are so fastened together that the prongs are at right angles to one another. These hooks are attached to the iron bar by strong cord or chains, usually chains, so as to be freely moveable. There are about four hooks on each chain. This contrivance is allowed to drag upon the bottom of the stream. As the mussel usually lies with its shells slightly opened whenever a hook strikes between them, the mussel shuts down on it and holds firmly until taken off. It is surprising how successful this method is and wherever it can be used it is very effective.

The best thing to open mussels with is an ordinary

steel table or case-knife. The blade is long and flexible and will follow the curves of the shell. There are two ligaments which hold the shells together. When these are cut the shells fall open, allowing an easy examination for pearls.

In opening the mussels there is one caution to be observed. It is this: Do not insert the knife at the tip or posterior end of the shell but always at the anterior (See Plates II and III). The reason is that the tip of the shell is always the brightest part and pearls formed there are usually the best. Therefore care should be used so as not to injure them with the knife.

Clams should not be opened while one is standing in the water or in any place where a pearl falling from the mussel would be lost. Valuable pearls have been lost by such carelessness.

In fishing for pearls, the mussels should not be opened by cooking or steaming, as heat is very liable to injure a pearl by destroying its lustre. It is said that at the time the Queen pearl was found in Notch Brook, New Jersey, a pearl weighing 400 grains and what would probably have been the finest pearl of modern times was rendered valueless by cooking the mussel. The practice of steaming or vatting the mussels is quite general with those who gather shells for the pearl-button factories. A false bottom is put in the vat to prevent any pearls falling from the shells from coming in contact with the heated bottom. Yet any heat injures a pearl, and dealers say that many fine pearls are injured or destroyed by this method of opening the mussels. Mr. John F. Morse, of Chicago, a large dealer in American pearls, declines to buy pearls that have the appearance of coming from mussels that have been steamed, vatted or cooked. When the mussel fishing is done primarily for the shells, the best method for removing the meats is by cooking. When the mussel fishing is done for pearls, it is advisable to open the mussels by hand.

Too great care cannot be used in searching the mussel for pearls. The best shaped pearls are usually lying in the mantle at the outer edge of the shell, and can be readily seen. Fine pearls are frequently enclosed in the meat of the mussel and this should be carefully felt over. The shells should be examined as pearls are often attached. If a pearl is found attached to the shell, it is better to send the shell as it is to the buyer who will be able to remove the pearl to the best advantage. Fine pearls are frequently found in discarded shells. It is said that the Queen Mary, one of the finest pearls ever found in the Mississippi region, was nearly lost by oversight. The clammer had thrown the shells aside, when his wife looking them over, discovered the pearl attached to a shell. The finders sold it for $1,700, and it is reported to have sold later for many times that amount.

Vane Simmonds, in his valuable book,[9] "Fresh Water Pearls," mentions the case of a boy at Genoa, Wis., who found a discarded shell to which was attached a large growth. The boy's father took the

---

[9] Fresh Water Pearls, by Vane Simmonds, Charles City, Iowa.

shell to a local jeweler who bought it for $2,675. When the incrustment was removed a fine saucer-shaped pearl weighing 186 grains was discovered. Its value was estimated at from $12,000 to $15,000.

The experience of pearl fishermen has shown that there are certain indications by which the presence of a pearl within the mussel may sometimes be de-, tected. It is occasionally indicated by a slight bump or curvature on the outside of the shell. Shells that are honey-combed or bear excrescences and are generally ill-looking, are most frequently found to contain pearls.

It is difficult to estimate about what per cent of the mussels on an average contain pearls. One estimate is as follows:[10] one pearl to one hundred mussels and one per cent of the pearls are of good quality. This is a conservative estimate and the actual percentage of pearls is probably greater. There is no certainty about it, and success in pearl fishing is largely a matter of good luck and hard work, principally the latter. Where mussels are plentiful as they are along most streams, a great many can be examined in a day. This makes the work profitable yielding on the average as much or more than ordinary labor, and there is always the possibility of making a rich find.

Many persons make pearl fishing a business during the season in which it can be carried on. There are many people living in the vicinity of streams who fish for pearls at odd times and when other work is

---

[10] Precious Stones, Max Bauer, p. 585.

not offered. Some do it as a vacation, camping along streams, enjoying the variety of the work and outing. Women and boys often make good money during the pearl fishing season. Anyone giving the work a fair trial is seldom unrewarded.

It is sometimes surprising that there is not more pearl fishing done by people living in the vicinity of streams. If a diamond field should be discovered, the excitement would be great. Yet the next most valuable gem can be found in every river, stream and lake in the country, but until someone reports a big find, only a few people give any thought to the resources before them.

It is not only the large rivers that have valuable pearl fisheries, but the small streams and creeks yield many of the finest pearls. Some of the streams are fished, some are already nearly exhausted, but the great pearl resources of the American streams as well as of lakes, remains as yet practically untouched. This wealth is the gift of nature, and needs only to be taken by those having the inclination and perseverance to do it.

VALUE.

In regard to the value of pearls much may be said in a general way though little can be written to aid one in determining the actual value of a particular pearl. There are so many circumstances affecting the value of this gem, that accurate judgment can be obtained only by one who has had a great deal of experience with pearls and who is in close touch with the conditions of the pearl trade. It is only possible for a writer to refer to those qualities which a gem pearl must possess.

Mr. Herman Myer has mentioned and described these qualities of a pearl in a simple and direct manner.[11] Mr. Myer has done very much to develop American pearl fisheries. He is a pearl dealer, thoroughly familiar with the pearl trade and has always encouraged pearl fishing, both by giving good information to those desiring it and by giving good money for pearls. Mr. Myer mentions the following qualities that must be taken into consideration in the order named in determining the value of a pearl: Brilliancy, Smoothness, Shape, Perfection, Size and Color.

BRILLIANCY. By brilliancy is meant the peculiar lustre of the gem. A pearl, to be of any value whatever, must be bright, like the brightest part of a shell.

---

[11] Pearls: How to Find and How to Value Them." Herman Myer, 41 and 43 Maiden Lane, New York City.

A mistake commonly made is to think that every concretion found within a mussel is a pearl and hence of value. If the pearl does not possess a luster it is of no value, whatever its size and shape may be. Pearls that at first appear to one as having a good lustre may prove to be quite dull when compared with really fine gems. A pearl possessing a brilliant lustre seems to shine of itself.

SMOOTHNESS. The surface of a pearl should be smooth, that is, free from wrinkles, knots, dents and the like. A slight imperfection of this sort may detract very much from the value of an otherwise perfect pearl.

SHAPE. The ideal shape of a pearl is round, though half-round (button pearls) and pear-shaped pearls have about equal value. Irregular shapes are of much less value, though the increased use of baroque pearls makes them of considerable importance.

PERFECTION. A gem pearl must be free from spots and blemishes. Anything that mars the beauty of a pearl spoils it as a gem. The demand for perfect pearls is very great because wealthy people are willing to pay large sums for such gems, while imperfect ones are not wanted. It is for this reason that a slight imperfection reduces the value of a pearl very much.

SIZE. Pearls are valued by weight, and the unit of weight is the pearl-grain, which equals ¼ carat. The price given is usually so much per grain flat. As the weight of a pearl increases its value increases

PLATE IV. NECKLACE OF FINE AMERICAN PEARLS.

This necklace numbers 41 pearls. There are 38 round pearls and three pear-shaped pendants. The total weight of the necklace is 1,935 grains. The largest pearl weighs 112 grains, the smallest weighs 24 grains. Three of the largest round pearls cost the dealer who made the collection $10,000 apiece. The necklace was sold abroad.

very rapidly. For example, jewelers frequently determine the value of pearls by so much per multiple grain, which means the grain number multiplied by itself and the product multiplied by the price per grain multiple. That is, at $3 per grain multiple, a 6 grain pearl would be worth $6 \times 6 = 36 \times 3 = 108$ dollars, while a 12 grain pearl would be worth $12 \times 12 = 144 \times 3 = 432$ dollars. Although the 12 grain pearl is only twice as large as the other, it is worth four times as much. When pearls are valued at a flat grain price the increased worth due to size is shown by a larger price per grain than would be paid per grain for a pearl of the same quality though of smaller size. Irregular shaped pearls do not increase greatly in value as they increase in size.

COLOR. White is the standard color for a pearl. Colored pearls frequently bring as much and sometimes more than white pearls of the same quality, but white is the color most preferred. Light pink, deep red, green and black are desired colors. Yellow, brown and dark pink are not. Besides a general preference for white pearls, another reason for the smaller value of colored pearls is that they are difficult to match and their use in the making of jewelry is limited.

There are outside circumstances that affect the value of pearls such as the fashion and condition of the pearl market. Moreover, a pearl is not like other articles of commerce in that as a usual thing there is no other pearl exactly like it and which will determine its value. Since there is no absolute standard to determine the value of a pearl, it must be fixed by agreement between the buyer and seller. As the demand for pearls is in the centers of wealth and fashion, it is usually for the advantage of pearl fishermen to sell their finds to responsible dealers who are familiar with the conditions of the pearl trade. Such buyers can usually do more for one in the way of price as well as in advising about pearls than persons who buy pearls for speculation.

There need be no hesitation about sending pearls to dealers. If the price they offer is not satisfactory, they will return the pearls safely. If it is so desired, some dealers will place a very fine pearl in the market and attempt to get the price asked by the person sending it to them.

DEALERS IN AMERICAN PEARLS.

It is not pretended that the following list of pearl dealers includes all who buy American pearls. Almost every manufacturing and retail jeweler is ready and willing to buy American pearls. The parties listed either deal exclusively in American pearls or at least make them a specialty, and have acquired a reputation for promptness and fair dealing. The names and business addresses of these dealers are given for the convenience of those who may care to patronize them.

Herman Myer, 41 and 43 Maiden Lane, New York.

Crossman Co., 3 Maiden Lane, New York.

L. Tannenbaum & Co., 15-17-19, Maiden Lane, New York.

John F. Morse, 700 Postal Telegraph Bldg., Chicago, Ill.

Max R. Green & Co., 301 Masonic Temple, Chicago, Ill.

S. J. Son, 103 State, Street, Chicago, Ill.

W. D. Burd, Vincennes, Ind.

Frank Koeckeritz, Vincennes, Ind.

William S. Miller, Clinton Iowa.

Arthur Reichman, Nassau Street, New York.

John M. Richard, 2 Maiden Lane, New York.

Maurice W. Grinberg, 527 Fifth Avenue, New York.

Alfred H. Smith & Co., 452 Fifth Avenue, New York.

Jos. Kaufman, 170 Broadway, New York.

Henry E. Oppenheimer & Co., 3 Maiden Lane, New York.

Amerinan Gem and Pearl Co., 14 Church St. New York.

I, Guntzburger & Son, 11 John St., New York.

Charles Adler's Sons, 527 Fifth Avenue, New York.

Bunde & Upmeyer Co., Milwaukee, Wis.; 68 Nassau St., New York.

Max Nathan Co., 68 and 70 Nassau St., New York.

CARE OF PEARLS.

Pearls are of a delicate nature and require the exercise of care in handling them. They should not be carried loose in such a way that they can rub or strike together.

Pearls should be shipped either by Registered Mail or by Express. It is not safe to send them by the ordinary mail as they are liable to be crushed in stamping or lost through the breaking of the package by the cancelling machine. When properly wrapt, they may be sent by Registered Mail or Express with perfect safety.

Pearls may be wrapt by folding them in strong linen paper and enclosing them in an ordinary envelope. If greater care is desired, they may be sent in small boxes made of stout paste-board or of wood.

It is not advisable to try to improve the appearance of pearls by artificial means. The lustre of a pearl cannot be made brighter by polishing or by any other process. A pearl is the work of Nature, and if it is not completed, it cannot be helped.

ORIGIN OF PEARLS.

It is evident that the formation of a pearl is not an ordinary or normal function of a mussel. If it were, all mussels or at least the greater number of them would contain pearls. As it is only the occasional mussel that produces a pearl, we are led to believe that pearl formation is started by some unusual or accidental occurrence from which most mussels escape.

There are various theories about this cause of pearl formation. These theories are not conflicting, however, and it will be seen that they are practically all based upon one principle, and that is the presence of something foreign to the mussel, which by its irritation or mere presence causes the mussel to secrete an abundance of pearly material in the form of a pearl. It is stated repeatedly, and there can be no doubt of the fact, that every pearl has a nucleus, that is, something in the center of the pearl which caused the mussel to cover it with nacre and thus create the pearl. Further, it seems evident, that there are many things which may act as nuclei or pearl-causing irritations, and so the formation of pearls may be started in several different ways.

Perhaps the most common theory about the cause of pearl formation is that a grain of sand or some other small inorganic particle, gets into the mussel and there becomes the nucleus of a pearl.

Closely related to this view, is the theory that pearls are caused by the eggs of the mussel which have failed to be expelled, and which act as nuclei for pearls. In 1826, Sir E. Home pronounced this to be the cause of pearl formation. The theory had been advanced somewhat earlier as the following letter written by Christophorus Sandius in 1763, and quoted by Home,[12] will show: "The pearl shells in Norway do breed in sweet waters; their shells are like mussels but larger. The fish is like an oyster, and it produceth a great cluster of eggs like those of crayfishes, some white, some black (which latter yet will become white the outer black being taken off); These eggs when ripe are caste out, but sometimes it happens that one or two of these eggs stick fast to the sides of the matrix, and are not voided by the rest. These are fed by the oyster against her will, and they do grow according to the length of time into pearls of different bignesses, and imprint a mark both on the fish and the shell."

It is no longer believed that the eggs of a mussel are the exclusive cause of pearl formation, but it is very possible that they may be one means of starting the growth of pearls.

Another view is that the formation of a pearl is started by an accidental injury to the mussel. This is based upon the fact that pearls are frequently found in distorted and deformed shells. It is thought by some that this injury or irritation may be created by certain forms of animal life that eat or bore into

---

[12] Phiolsophical Transactions R. S. (1826), Vol. 2, page 338.

the shell, causing the mussel to secrete narce to protect itself, and in this way occasionally forming pearls.

Another theory is that pearl formation is started by a disease of the mussel which acts as a source of irritation. This view is based upon the fact that a large number of pearls are found in certain streams or at certain times indicating that many of the mussels might be affected by some disease that acted as a pearl producing cause.

A recent theory and one founded upon careful scientific observation, is that pearl formation is caused by parasites. These are minute worms or other very small forms of animal life that get within the oyster and become the nuclei for pearls. This theory has been advocated during recent years by scientific men, especially those who have been working upon the subject at the request of governments owning large marine pearl fisheries. Prof. W. A. Herdman, Mr. James Hornell, Dr. Jameson, M. Diguet, M. Dubois, M. Seurat and others have contributed much to the scientific knowledge of the pearl oyster and its pearl. The conclusion that seems to have been reached at present is that while parasites are not the only cause of pearl formation, they are the principal cause in producing the fine pearls.

Prof. Herdman, who has been working at the British pearl fisheries in Ceylon, says,[13]—"Turning now to the subject of pearl formation, which is evidently an unhealthy and abnormal process, we find

---

[13] Popular Science Monthly, Vol. 63, page 235.

that in the Ceylon oyster there are several distinct causes that lead to the production of pearls. Some pearls or pearly excrescences on the interior of the shell are due to the irritation caused by boring sponges and burrowing worms. Minute grains of sand and other foreign bodies gaining access to the body inside the shell, which are popularly supposed to form the nuclei of pearls, only do so, in our experience, under exceptional circumstances." Prof. Herdman then concludes that most of the pearls found free in the tissues of the Ceylon oyster are caused by certain parasites.

It seems plausible to believe that many of the fine pearls are formed by parasites. The parasitic nuclei are so small that pearls of perfect shape would be likely to occur, and it is possible that parasites would be more apt to affect the portions of the oyster where free and perfect pearls are formed, than inorganic particles which the oyster could rid itself of with greater ease.

It by no means follows, however, that all pearls or even all fine pearls are caused by parasites. As suggested before, there are no doubt several different ways by which pearl formation may be started. Moreover the theory that a grain of sand or some other small particle becomes the nucleus of a pearl, has the evidence of actual demonstration. Fine pearls have been found to be formed over pieces of clay and other particles of considerable size. Even tiny crayfish and other small animals have been found encrusted by nacre. Besides this, the experience of the Japanese in growing culture pearls demonstrates that pearls may be formed over foreign particles as nuclei.

There are even more positive proofs that the nuclei of pearls are very frequently inorganic particles. "Year by year some thousands of pearls are cut in half by working jewelers, and their universal experience is that a nucleus is always to be found."[14] A pearl cutter and driller has this to say:[15] "The halving of a globular pearl is naturally through its center. Now, I have carefully examined the *wrong* side of thousands of half-pearls, noting the concentric rings or lines which indicate the onion-like construction: using a powerful loupe, it is found usually that the center is a speck or something other than solidified nacre; always a speck of material which is not identical with its surroundings. Picking at them with a fine-pointed needle loosened them, and under a strong glass it could be discerned that they were specks of mineral, vegetable or other composite matter which was the nucleus around which the nacreous matter was deposited. Valuable pearls have been frequently rendered useless for their purpose by the ignorance of the driller that there is surely a central core, mayhap a speck of hard mineral in the linear course of the drill." Sometimes the center of a pearl is a cavity indicating that the nucleus may have been a form of animal life which had perished, leaving its shape impressed in the center.

It is probable that the majority of the fresh-water

---

[14]Pearls and Pearling Life, by Edwin W. Streeter, page 109.
[15]Jewelers' Circular Weekly, July 26, 1905.

pearls are formed from small inorganic particles as nuclei. Every stream carries a great deal of sand and particles of mineral in suspension and the bed of the stream on which the mussels lie is always in commotion. As the mussel draws a current of this water through its body in feeding, it must take in myriads of these small substances. It is natural to conclude that many of these act as nuclei for pearls. The abundance of irregular shaped fresh-water pearls tends to support this conclusion.

CULTURE PEARLS.

"Among the picturesque industrial possibilities of our southern Pacific coast is the articfiial production of pearls. By this is meant, not the manufacture of artificial pearls, but the artificial growing of real pearls; that is, instead of the haphazard pearl fishing of the present, the establishment, on the Southern California coast, of oyster ranches, where the pearl-producing bivalves shall be scientifically directed and assisted in growing both gem pearls and mother of pearl."[16]

The writer of the article referred to above suggests a matter of considerable importance and one that has received the attention and exercised the ingenuity of many persons. The search for pearls is uncertain, and at best the production of fine pearls is exceedingly small. If a method could be devised by which the formation of a pearl could be started, it would then be carried on and completed in the same manner as natural pearls are formed. Moreover, the beginning of the pearl formation not only could be made certain by such a method, but it might be guided and directed to the end of securing a greater number of perfect pearls.

The theories of pearl formation have already been considered. In this connection it was seen that all the theories were based upon the intrusion or pres-

---

[16]Popular Science Monthly, 49:390.

ence of something foreign to the mussel. That this foreign particle, be it parasite, egg of the mussel, grain of sand or something else, acts as a nucleus for the pearl, receiving successive coatings of nacre. Working upon this principle many attempts have been made to grow pearls by introducing various foreign substances into the oyster to induce the formation of pearls. A brief review will be given of some of these attempts to grow pearls, the methods used, and the success with which they have been attended.

It is a well known fact that for centuries the Chinese have been accustomed to insert small figures or images of the deity Buddha between the mantle and shells of river mussels. These figures are usually made of tin, tinfoil, lead or some metallic substance. These are allowed to remain in the mussel a few months in which time they are covered over with nacre. They are always attached to the shell and have to be cut from it. These images are curious, but their commercial value is practically nothing.

It is to be observed that the object to be obtained is to induce the formation of "free" pearls, that is, pearls that are not attached to the shell. It is comparatively easy to secure the covering of images with pearl by a method similar to that used by the Chinese. The difficulty is to obtain free pearls of such quality as to be of commercial value.

In the year 1761, Linneaus, a famous Swedish naturalist, made known that he had discovered a method by which mussels might be made to produce pearls. He sold the secret to a company that desired

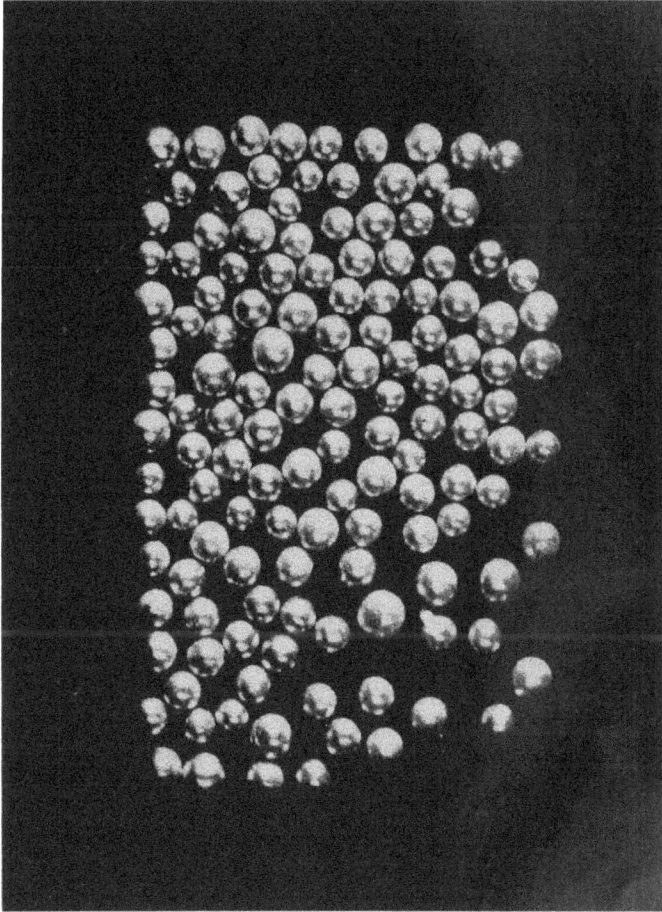

PLATE V.   JAPANESE CULTURE PEARLS.   NATURAL SIZE.
(From Bulletin United States Bureau of Fisheries for 1904, Plate XI.)

to try the method for its commercial value. Little seems to be known of the method or of its practical success. It has been said to have consisted of making a perforation in the shell, without the introduction of any foreign substance.[17]

Experiments to cause freshwater mussels to produce pearls have been undertaken at the pearl fisheries of Saxony. A foreign substance is either introduced into the mantle to form the nucleus of a free pearl, or placed between the mantle and shell. The following quotation shows that the experiments have been attended with some success. "The cultivation of the pearls of fresh water mussels has become an industry of considerable importance in Saxony and other parts of Germany. The pearls are generally inferior but occasionally a gem of real excellence is produced."[18]

In recent years the Japanese have had remarkable success in growing pearls. The following paragraph by K. Mitsukuri, Professor of Zoology in the Imperial University of Tokyo, is from the Bulletin of the Bureau of Fisheries, 1904, page 283: "In 1890 I suggested to Mr. Mikimoto, a native of Shima, who had grown up and lived in the midst of the pearl-producing district, the desirability of cultivating the pearl oyster. He took up the subject eagerly and began making experiments on it. Soon after I pointed out to him also the possibility of making the pearl oyster produce pearls by giving artificial stimuli. He

---

[17]Notes and Queries. 7th Series, Vol. I, p. 128; Vol. VI, p. 125.
[18]Popular Science Monthly, 25 :430.

at once proceeded to experiment on it. The results have been beyond expectations, and today the Mikimoto pearl-oyster farm, put on a commercial basis, has thousands of pearl-oysters living on its culture grounds, and is able to place annually a large crop of pearls on the market."

This company has been given a concession by the Imperial Government respecting rights to the sea in the bay of Argo. Here the pearl-producing oysters are grown and cared for. At the proper time they are treated to induce pearl formation. The methods known to be used, consist of the introduction of nuclei to be covered over with nacre. The nuclei are pieces of pearl shell cut in the shape of high button pearls. These are placed between the mantle and shell of the oyster, and allowed to remain from one to four years. When they are removed, they are usually attached to the shell, and so are in the shape of half pearls or a little more than half pearls, "but as regards lustre, shape, and size, they are beautiful beyond expectations, and meet the requirements completely in cases where only half pearls are needed."[19]

It is evident that the pearls produced were of good quality. It is said that when Mr. Mikimoto submitted twenty-seven of the pearls to some royal people of Japan, they were so pleased with the gems that they bought them all at a large price.[20]

If the theory that most pearls are formed by the presence of parasites is accepted, other methods to

---

[19]Bulletin of the Bureau of Fisheries, 1904, page 284.

[20]The Jewelers' Circular Weekly, Sept. 19, 1900.

induce pearl formation are required. In an article, "On the Origin of Pearls,"[21] Dr. H. Lyster Jameson concludes that fine pearls are caused by a parasitic worm, and says: "As an economic result of these investigations, it would seem that the artificial production of marketable pearls in large quantities should present no great difficulties, if the particular cases be intelligently investigated. The fact that trematodes have been ascertained to be at least one cause of pearl formation in several of the mollusks that produce marketable gems gives us every reason to hope that, by learning the life-histories of these parasites, we may be able to infect any number of pearl-growing oysters or pearl-mussels to any desired extent, without any operation on the individual mollusks, by simply placing them in the proper surroundings, in company with infected example of the first host. Once infected, the mollusks could be bedded out on suitable grounds, and left to care for themselves, until the pearls formed in them were of marketable size."

This is well in theory but it has not as yet been worked out in practice. It is said that M. Dubois, a French naturalist, has succeeded in communicating the parasitic infection to oysters in such a way as to produce one or more pearls from every ten oysters. The pearls, however, were not of a size or quality to be of value.

A company has recently been formed to grow

[21]Proceedings of the Zoological Society of London, 1902, Vol. I, pp. 140-166.

pearls at the Gulf of California. Its method includes the incubation and hatching of the pearl oysters, and a process of innoculation to induce pearl formation. The company has extensive concessions for pearl fishing, and it is not known exactly what success has been had with cultural pearls.[22]

Referring to the work of the Chinese, Mr. George F. Kunz says:[23] "This method of producing figures and symbols that could be used for ornaments is one that would recompense any American who would produce the same results in some of our richly colored and brilliantly lined Unios." The writer of this article is aware of but one such enterprise in the United States, engaged also in the production of culture pearls, and with apparent success. In 1906 it was reported that the production of culture pearls in America was a rapidly developing industry, but there seems to have been little foundation for the report.[24] It is not impossible that mussel farming may become an industry of some importance in the near future.

"Pearl-oyster culture is still in its infancy, but its promises are bright. If, in addition to half pearls, full or 'free' pearls can be produced at will, as there are some hopes, it will be a great triumph for applied zoology."[25]

[22]The Jewelers' Circular Weekly, Nov. 18, 1903.

[23] Bulletin of the United States Fish Commission for 1893, p. 456.

[24] The Jewelers' Circular Weekly, Aug. 1, 1906.

[25] Bulletin of the Bureau of Fisheries, 1904, page 284.

www.ingramcontent.com/pod-product-compliance
Lightning Source LLC
Chambersburg PA
CBHW022342280326
41934CB00006B/745